Protecting Your Castle

A Common Sense Guide for Home Safety and Defense

Brett Lechtenberg

COPYRIGHT

Copyright © 2015 by Brett Lechtenberg and The Family Success Network LLC.

All rights reserved. No part of this book may be reproduced or transmitted in any form or by any means, electronic or mechanical, including photocopying, recording, or by any information storage and retrieval system, without permission in writing from the publisher.

Published by Brett Lechtenberg and The Family Success Network, Sandy Utah.

ISBN-9781532907012

IMPORTANT NOTICE - DISCLAIMER

While we believe that the Protecting Your Castle Safety Training information and programs provided by Family Success Network LLC are very effective, there are inherent risks to anyone in situations involving family protection and home safety, such as physical, mental, and emotional violence and even death.

The information in this document and associated videos is an adjunct to, but not a substitute for, the advice of physicians, psychologists, counselors, and law enforcement. The reader should regularly consult these professionals in matters relating to the mental and physical health of their children or family members and particularly with respect to any symptoms that may require diagnosis or medical attention.

Users of Family Success Network LLC and Brett Lechtenberg's materials acknowledge that the responsibility for their personal or children's safety lies solely with them and limit the liability of Family Success Network LLC and Brett Lechtenberg to the purchase price of the information and programs.

DEDICATION

This book is dedicated to all home owners around the world who have to worry about having their home or possessions put at risk by criminals.

Let's all do our part to make our homes and neighborhoods a safer place.

ACKNOWLEDGEMENTS

This book and program would not have been possible without the help and support of so many people. First and foremost, thanks to my wife Teresa and my children, Dalton and Carter, who gave me the time and encouragement to complete this project.

Special thanks to my incredible staff who continued to work hard and take on extra duties and allowed me to finish this project: Brittany Alleman Ayers, and to the parents and students of our facilities who continually gave me feedback and follow up while we tested our program.

Professional thanks to my friends Dennis Bullard, , David Ventura, Sal Rossano and Dave and Jodie Osofsky for their continued willingness to review material, layout websites, read copy, look at photos, set up entities and everything else that it took to produce this program.

A big thank you to my colleagues and trainers throughout the years who have dedicated their lives to the betterment of society by creating safety programs, teaching self defense, studying survival and always being willing to share their vast amount of expertise. Thank you Richard Na, Tom Patire, Thomas Fischer, Russell Wright, John Nottingham and the rest. I am truly lucky to have all of these people in my life.

TABLE OF CONTENTS

Chapter 1
Introduction 1

Chapter 2
Understanding the Criminals We Are Dealing With 5

Chapter 3
How and Where Most Home Burglaries Happen 11

Chapter 4
Protecting Your Home through Security Assessments, Burglar
Proofing, and Staging 23

Chapter 5
What if You Find Someone in Your Home?........................ 41

Chapter 5
Wrapping it All Up 51

Appendix 53

References..63

About the Author 65

CHAPTER 1

Introduction

Welcome to *Protecting Your Castle*, a common sense guide to home safety and defense. I want to take a moment and say thank you for purchasing this program and tell you about why it came into being.

Why Did This Program Come Into Being?

Everybody's home is their castle, and I believe you should feel safe and secure in your castle, and also that you have the right to do as you please and follow your dreams in your castle. Unfortunately, bad things happen to good people all the time. I want to help you protect your family's safety and your dreams.

My mission is simple: I want to educate and empower people like you on how to constantly improve their personal and family safety. I want to help you feel better about everything you do in your home and know your family is safe. I want to give you concepts and ideas that you can put into effect in less than 30 minutes after finishing this program.

My name is Brett Lechtenberg. I am a family empowerment and safety expert. As always, if you have any questions with this, or any of our programs, feel free to contact me at brett@brettlechtenberg.com or find me on Facebook.

What We Will Cover in This Program

Now, here is what we are going to cover in this program:

- The difference between a burglary, a robbery, and a home invasion

- When most break-ins happen

- The most common entry points for a home break-in

- The times that most break-ins occur

- What a criminal timeline is and how that works

- The most commonly stolen items in a home burglary

- Which rooms are targeted in which order and why

Protecting Your Castle

- How to do a security assessment of your home and see what a burglar sees

- My top tips on the quickest and cheapest ways to deter break-ins

- Having an action plan if someone breaks into your house while you are home

- What to do if you confront someone who has broken into your home

- "Security staging" your house for your protection

Tips for Maximizing This Program

Make the time to read this program in its entirety.

Upon completion of reading the program, give yourself a timeline of completing as many of the tips that fit your situation as you possibly can within the next 72 hours. Then, go back and reassess your home security again in 7 days and make any upgrades you may have missed.

Finally, make sure each member of your family is involved and educated about home security, and yes that means the kids too.

Brett Lechtenberg

SUMMARY

- Complete the program within 72 hours
- Get the entire family involved in the training

CHAPTER 2

Understanding the Criminals We Are Dealing With

Understanding the Difference between a Burglar, Robber, and Home Invader

In this section I want to make sure I take some time to explain the differences between the average burglar, a robber, and a home invader. Many times, people use these terms interchangeably and that is a dramatic mistake. They are three very distinct types of criminals.

Burglar

A burglar is a person who enters a home, car, or other structure with the intent of committing a crime. Usually they want to steal something and escape without making contact with another person. They want to get in unnoticed, take what they want, and get out quickly.

Robber

A robber on the other hand is a criminal who enters a home, car, or other structure with the intent of committing a crime and they are prepared to use violence. Usually they want to steal something and they either know or assume that there will be people present. They want to get in unnoticed, take what they want, and get out quickly; but they are very prepared to use violence.

Home Invader

Finally, we have a home invader. A home invader is the worst type of person. This is a person who enters a home, or other structure, with the intent of committing a crime and using extreme violence. These people are true predators. A true home invader does not value human life the way a normal person does. They want to get in unnoticed and take what they want, but they are prepared to take their time and abuse you and your family for as long as they want.

What This Course Is About

This course is about dealing with break-ins and non-violent burglars. I am going to give you some tips for dealing with robbers and home invaders, but more information on dealing with physical force check out my other programs.

Protecting Your Castle

A Burglar's Mentality and Timelines

I want to take some time in this section to talk about two main concepts: one being a burglar's mentality and the other being a burglar's timelines.

Burglar's Mentality

First, burglars want to be in and out of your house as quickly as possible. They are not going to sit in there and loiter. Second, they want no contact with people if they can absolutely avoid it. Third, they are not interested in violence. If they were interested in violence they would break into your home at different times than they do, which leads us right into the timeline discussion.

Burglar's Timelines

The first timeline is time of year. The biggest time of year for burglars is July and August because they know that people are inherently out of their homes, enjoying vacations, exploring the world, etc. They know that a lot of times, people have got their whole family with them and they are out doing other things.

Burglaries also go up during the holidays because people are more desperate, or on the flip side, if you are the person who has been burglarized, it may because you have done something foolish such as leaving things exposed in your car or leaving them somewhere they are an easy opportunity for someone to get at.

The second timeline is time of day. Since burglars are not interested in interacting with people and using violence, they are going to break in at a time of day when most people are not home. Most people are not home between 6 am and 6 pm, with prime time being 10 am and 3 pm. Even people who do not work are generally out between 10 am and 3 pm taking care of things of things for their family, for their spouse, etc. and they try to be back by the time their kids get home from school or their spouse gets home from work.

The final timeline is the time that a burglar spends in your house. They are only going to be in your home for an average of 8 to 12 minutes. They do not want to be in there a long time and have to worry about someone coming home or being noticed and getting caught.

Many times after a burglary, people feel very violated. That is natural. They think that someone has been in their house, going through all their stuff, touching all their possessions, and doing heaven knows what with them. The reality though is that does not happen much. A burglar wants to get in steal specific items from specific rooms that have the highest probability of holding something valuable, and then they are going to be gone.

So, take this information as you think about your day and as you are planning your security around your home to keep your family safe and your possessions safe and put it to good use.

SUMMARY

3 Types of criminal who break into your home

1. Burglar

2. Robber

3. Home Invader

Understand a criminal's timelines

Understand a robber's mentality

Brett Lechtenberg

Protecting Your Castle

CHAPTER 3

How and Where Most Home Burglaries Happen

Areas of Break-In

The Front Door

This is where most home burglaries originate from. It is the most vulnerable place on your home. About 35% of all home burglaries happen right at the front door because people don't lock their front doors and anyone could walk right in. It is really important, and it sounds overly simplistic, but 35% of the time you can stop a home burglary by simply locking your front door. Having a key pad is great and always using a deadbolt is awesome. You want to make sure that you lock your front door.

In the town that I live in, there were 18 home break-ins last month. 17 of those came about because someone did not lock their door. Think about that.

First Story Windows

If you want to stop at least 25% of all home break-ins, lock your first story windows. Just like the front door, people think they are safe because nobody is going to bother to walk in, nobody would dare, but people do because they know that it is a target of opportunity and it is easy to take advantage of. So lock those first story windows; keep yourself and your family safer.

Side Doors and Back Doors

If you want to cut the chance of your home being broken into by about 22%, then close and lock your side entrance and back doors. Use the deadbolt, not just the handle. Assume the handle is unlocked; keep yourself and your family safer.

Garage Doors

Here is a way to cut another 9% off the potential for a home burglary. Lock your garage door. Close those doors completely and lock them up. Many times a burglar will see an object of opportunity (a bicycle, a compressor, tools, etc.) and they will go in and take those things and you won't even know they are gone until you go looking for them to do some sort of project, ride around the neighborhood, etc. You won't even know. So lock the garage door. Keep the doors closed when you are at home, even if you might be going in and out. I understand that it might be a bit of an inconvenience, but people can steal things very fast.

Protecting Your Castle

It is also important to note that even when garage doors are partially open you are vulnerable. So close them all the way and lock them up.

Second Story Windows

The last place that a burglar is going to break in is at a second story window. Only about 2-3% of all break-ins happen on second story windows; however, it is worth noting. Make sure you keep them locked. It is also a smart idea to keep a piece of doweling inside them, especially if you are not home for a long period of time.

Be especially conscious of these windows if they are bordering your backyard, a wooded area, or somewhere else where people don't normally go or can't normally see your back windows. So, food for thought, take it seriously.

Most Burglarized Rooms

Master Bedroom

The #1 most burglarized room in your home is the master bedroom. If a burglar had to break in and pick just one room that they were going to go to, generally they are going to come here. They know that the things people value most are the things they tend to keep the closest to them; and the most personal room in a home is typically the master bedroom.

People are going to have things like jewelry boxes that hold things they don't wear very often (fancy necklaces, rings, etc.) and the things that they wear every day are also going to be kept close by (wedding rings, watches, etc.).

People are also going to have personal things like prescription drugs in their master bedroom. Many times people have to take medications at different times of day or night and having it right next to the bed makes it very convenient.

Finally, cash. Cash is one of those things that when people do not keep it in their wallet, they will find a safe place in their bedroom to stash it and keep it in case of an emergency. That is why a burglar will generally pick the master bedroom as the #1 spot to burglarize in a home.

So, think about that as you are going through your house doing your security assessment. Think about how you can place things differently (maybe in a safe or somewhere else inside your house) to keep your possessions safe somewhere outside your master bedroom.

Home Office

Typically the second most burglarized room in a home is going to be the home office. Home offices have all kinds of great electronic devices (laptop computers, printers, scanners, paper shredders, desktop computers, etc.). All of these things are pretty easy to take, they are not very fragile, and

Protecting Your Castle

people generally do not have the serial numbers written down or any distinguishing marks on them. All of these things make them really easy to sell on the open market or to pawn. In addition, a home office is generally going to have checkbooks, receipts, bank statements and a lot of other things that could be used to steal someone's identity.

So, you want to make sure that in your home office you have good inventory of everything. Also, get those serial numbers, put some distinguishing marks on your items and make sure that you keep any old bank statement or credit card statement that you don't need either locked up or shredded.

Master Bathroom

Another one of the most burglarized rooms in the home is the master bathroom. People keep a lot of valuables in their master bathroom, because they see it as a very safe place. They will have things like a jewelry box, where they might have their wedding ring, expensive jewelry, necklaces, earrings, watches, etc.

So if you have really expensive things, heirlooms, etc. that you want to keep safe, then do not keep them in the main jewelry box in your house. Hide them somewhere else. Put just enough in your main jewelry box that if someone breaks into your house it would satisfy them but they would think, "Hmm this person doesn't have very much," or, "They don't have much here now," and they are more inclined to move on their way. Keep those really valuable things (family heirlooms, etc.) in a safety deposit box at the bank or in a safe somewhere else in the house. I know it sounds like a little bit of an inconvenience, but those things that you value most will be a lot safer.

Living Room

Typically, the fourth most burglarized room in a home is the living room. There are a lot of valuable things here: artwork, family heirlooms, stereos, television sets, DVD players, etc. All of those things have a good value, can be sold quickly, and burglars know many times that people are not very good at writing down serial numbers and model numbers of their possessions. That makes these items pretty easy to move, they can sell them at pawn shops, etc.

So make sure that when you get something new make it a habit to fill out the warranty, write down the serial number, write down the model number, take a picture of it, and then keep those things in a safe place. Then, should you have to report a home burglary, you have all the information necessary to give to the insurance company, to give to the police, and get your possessions back if at all possible – or at least get full claim for the items from your insurance company.

Formal Dining Room

Rounding out the list for the most burglarized rooms in your home is the formal dining room. Of course in these rooms people have things like crystal, china,

Protecting Your Castle

fancy silverware, heirlooms, other collectibles – and these things have a definite high value sometimes. However, they are a little less likely to be stolen because the average burglar is not going to be in your house long enough to deal with them properly. By that I mean these things have to be in one piece and they have to be in good condition in order to be sold on the open market and have real value. Because the burglar is not going to be in the house very long, they typically do not have the time to deal with these things, package them, etc. so that they can take them out and keep them safe.

However, it is still a great idea to make sure that you have a good inventory of the things that are in your dining room. Make sure that you take pictures of these items so that if they do get stolen or vandalized you can report to the insurance company and you can report to the local police.

Commonly Stolen Items

In this section I am going to go over the most commonly stolen items during a home burglary.

Cash

As you might have guessed, cash is king. It is the number one most stolen item. It is important to understand that if you keep large amounts of cash in your home for some reason, most insurance companies will not reimburse you for amounts over one hundred dollars.

Make sure you check your policy, but many policies have this written into them.

Electronic Devices

The second most stolen item is electronic devices. Things like laptops, tablets and smart phones are easy to carry, expensive and they are easy to sell. Additionally, most people don't have the serial numbers written down for their items and they don't bother to put on any identifying marks.

Gold and Silver

The third most commonly stolen item is gold and silver. Jewelry, bars, coins, etc. are really easy to sell, they are really easy to conceal, and they have a big value on the open market.

Guns

Fourth is guns. Like it or not, we live in a gun culture and they are really popular. They are easy to carry, they are easy to conceal, and they are really easy to sell. So be careful with your guns. Make sure you lock them, put them in a safe, and keep them somewhere where they cannot be found.

Assorted Watches and Jewelry

Fifth is assorted watches and jewelry. A lot of people keep things in their jewelry boxes like necklaces, rings and other things that are very valuable, can be taken easily and sold quickly. Make sure that your really precious personal belongings are kept

Protecting Your Castle

separately from your average everyday jewelry. If something is really valuable, then lock it up in a safety deposit box at the bank.

Prescription Drugs

And finally, prescription drugs. After you are done with a prescription make sure you properly dispose of your prescription drugs at your local police station or pharmacy. Call your local police station to find out where the best disposal places for prescription drugs are.

Brett Lechtenberg

SUMMARY

Most Common Areas of Break-In

- Front door

- First floor window

- Back and side doors

- Garage doors

- Second floor windows

Most Commonly Targeted Rooms

- Master bedroom

- Home office

- Master bathroom

- Living room

- Formal dining room

Most Commonly Stolen Items

- Cash

- Electronics

Protecting Your Castle

- Gold and silver

- Guns

- Prescription drugs

Brett Lechtenberg

Protecting Your Castle

CHAPTER 4

Protecting Your Home through Security Assessments, Burglar Proofing, and Staging

Security Assessments

In this section I will tell how to do a security assessment of your home and make your home, your family and your possessions a lot safer.

The Concentric Rings of Protection

First I want to go over the "Concentric Rings of Protection." I am going to walk you through a diagram and some concepts that many of my clients have found to be very beneficial when setting up their family protection plans. Think about it as stripping away the layers of an onion.

The first thing you are going to look at is your outer yard and fence line. Next are the main doors and windows. Then I will go over the seldom used rooms in your home, followed by the sleeping quarters, and finally I will tell you a little bit about safe rooms.

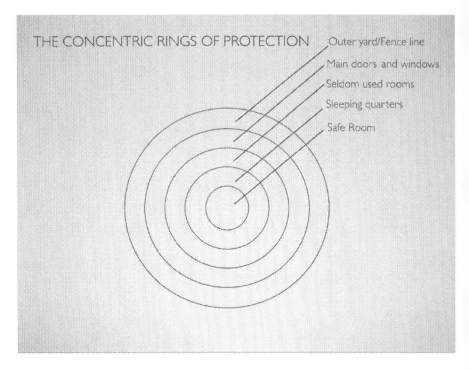

Outer Yard and Fence Line

First is the outer yard and fence line. The simplest thing to do is walk across the street from your home, turn around, and look at your house. Look at vulnerable spots. Think about things like:

- Are my windows and doors that cannot be seen from the street, secure?

- Are they easy to get to?

- Are my window wells inside my fence line?

Protecting Your Castle

- Are there thorny bushes and flowers underneath the windows?

These are all things that a burglar looks at to evaluate a home and tell whether or not that house is easier to break into than another house.

Then look for items left in your yard that could actually help facilitate a break in. Burglars are looking for easy opportunities wherever they can find them. Do you have an unsecured ladder outside? Do you have tools left in the yard from some project you were working on? Any of these things could help a home burglar break into your house.

Main Doors and Windows

Next are the main doors and windows. This is where most break-ins are going to occur. 35% of break-ins occur at the front door because the door is not locked. Another 22% of break-ins occur at the side doors and back doors because those are not locked. And finally, 23% of break-ins occur at the first floor windows because those are also not locked. So ask yourself, do I close doors and windows when they are not in use? Do I keep them locked? And do I maintain them in good repair? Do I have motion sensor lights on these areas at night? If you don't, these are simple things that you can do, with very little money, to make yourself and your family a whole lot safer.

Seldom Used Rooms

Next are the seldom used rooms in your home. Many people have things like storage rooms and unfinished basements. These rooms are rarely occupied and they are great places for people to break into because they can get in unnoticed. In addition, because they know they are not occupied and that you are rarely going to come down there, it can give them more time in your house to look through your stuff and come up with their own plans of action.

Keep any identity related information in these areas locked up and secure. Don't just assume that because these things are in your house they are secure. Make sure you put them in a safe or some other area that has a lock and key, and then make sure the rooms themselves are secure.

Sleeping Quarters

Next is the sleeping quarters. This is something that everyone really needs to focus on to make sure that they have good plans of action here. Whether it's your bedroom or your kids' bedroom, make sure you have good locks on all these rooms. Make sure you know the proximity to stairways, doorways, and other exits – not just for home burglaries, but think for example if you had a fire. Would you be able to get to your kids quickly? Would you be able to get out of the house? Would you be safe going down your own hallway and stairway? Would there be any toys, clothing, or other barriers on the stairs that could cause you or anyone else to fall, get injured, and then never be able to get out of your home? You've got to think clearly about these things. Are there ladders for escape from second story windows? So many people do not have these basic security measures and they are putting themselves and their family at risk.

Safe Rooms

The final area to talk about is safe rooms, and no I do not mean something like you see in some Hollywood movie. Safe rooms are generally going to be something like a master bathroom or a master bedroom. My favorite is a master bathroom and something that is near the street where you can be seen or heard should you need to yell for help.

I like bathrooms because the doors usually already have locks on them, they have got running water should you ever need it, they usually have some type of medical supplies (even if you don't have a full medical kit, you generally have rags, band aids, and some other things), by code in most places they have to have windows big enough to escape out of, they have toilets should you be trapped in there for a long period of time, and finally they are good place to charge a cell phone. Maybe have your spouse charge their cell phone next to the bed and you charge yours in the master bathroom. If you get stuck in your master bathroom and you are using it as a safe zone, then you will have at least one phone that you can call out and try to get help.

Burglar Proofing Tips

Now I want to share my top tips for burglar proofing your home. These are my top tips because they are either totally free or they are really cheap, they are super fast and virtually anyone can do them. Of course there are a lot more complicated things you could do with alarms, security systems, etc. Feel free to work those into your budget when you have time, you have the money and you can put those into place. But for now, here are some things are really cheap and easy to do.

Top Tips for Burglar Proofing Your Home

First and foremost, always lock your doors and windows. Also be sure to close and lock your garage doors and windows. Next, use your deadbolts and not just handle locks on your doors for that little extra protection. Also, replace the standard screws that are generally used for installing deadbolts with extra long screws that are at least three inches long.

Make sure that you have line of sight to everywhere in your yard. Be sure to keep bushes trimmed down and trees trimmed up three to four feet so nobody can hide behind them. Make sure there is no easy access to your backyard whenever possible. If at all possible do not put handles and hinges to fences on the outside. Keep them on the inside of your fence so it is harder to tell where the access points to your yard are. Also, whenever possible build your home

Protecting Your Castle

fence so that window wells to your basement are on the inside.

If you have a garage door with access to the backyard, disable the handle from the garage side. Then nobody can break into your garage and have easy access to your backyard. However, keep the handle on the backyard side usable so you can get in and out of your garage. Next, keep your outside lights on at night. Simply use LED lights to keep the cost down.

Plant thorny shrubs and flowers under any windows that are easy access points into your home.

Place a security / alarm company sign in your yard, even if you don't have a security system.

Place dog bowls on the floor where they can be seen from the entrances to your home, even if you do not have a dog. This has been proven to keep out intruders.

Keep door stops next to each entry door. This will make it difficult for someone to break in and get past you if you answer the door and they try to push by you.

Be sure to used doweling, metal rods or 2x4's to secure sliding doors. You can use that same type of doweling, metal rods and even nails hammered in at a 45 degree angle to keep windows, especially on the main floor, closed and secure.

Vulnerable Areas

Back Doors and Side Entrances

This is another area of opportunity for a home invader. These types of doors that enter into a basement or maybe a side entrance of a house can be very vulnerable. We don't usually spend a lot of time in our basements late at night because we are usually up in bed asleep, so home invaders are going to take advantage of that. On doors like this you want to use the deadbolt for sure. Make sure that there is some sort of 2x4, doweling or something in there that can be used as a doorstop. If you have sliding doors, put those things right in the tracks. If you have doors that open inwards, you want to make sure that you have a doorstop in there. There are lots of options that you have, but you really have to think about it.

And finally, you need to think about the glass. With doors like this one, go the extra mile and make sure that it is a really strong type of tempered glass or plexi-glass that is designed to handle a lot of punishment. There are a lot of great options out there, so make sure that you do your homework.

Big Basement Windows and Window Wells

This is another easy spot for a potential invader to come into your home. A big basement window and/or a big window well makes it so that once the home invader got down inside, it is so deep no one would see them from outside. Even someone standing in the yard just a couple of feet away might not see or hear someone inside a window well. Many times these windows get left unlocked because people don't think about it. They don't think about using a piece of doweling or a 2x4 to keep the window from opening. In addition, many times people don't put a grate on the window well opening. These are the type of things that you need to be aware of, not that you cannot have big basement windows and window wells, but that you need to secure them properly.

Staging for Your Protection

Next I am going to tell you about weapons of opportunity, which are things that you can use in your home to stage for your family's protection.

Right off the bat I want to say that I have seen people come up with some really extravagant things for home protection that just do not make a lot of sense to me. I have seen everything from having wall safes behind pictures for guns to stashing guns in special places under counters, etc. I don't know about you, but I don't want that many guns floating around my house. I have nothing against guns. But how many of those do you want to buy? How many do you want to have sitting around?

I have seen people talk about having nylon knives or other types of other nylon weapons that can be purchased on the internet or at gun shows. Those things are great if you want to go there, but they are expensive, and they don't give you any more advantage than the stuff that I am going to talk about. So if you want to go that route, go ahead; but I am going to tell you about things that blend into the background, are really inexpensive and are highly, highly effective.

Screw Drivers

First, screw drivers. A good screw driver will replace that nylon knife anytime and screw drivers blend into the background. You can have a lot of them around the house and they are relatively safe until you try to use them in an inappropriate manner. You can stash those screw drivers all over the place and a home invader or someone else is just going to look at those and go, "Oh this person left a screw driver there during their last home fix-it project," but it blends into the background enough that it's a secondary thought. Most people aren't even going to see them because they blend in so well, so think about that.

Hornet Spray

Hornet spray is a great one to replace the myth of pepper spray. Pepper spray can be great if you are trained on how to use it, you really know what it is about and you have really practiced with it.

The unfortunate thing is most people do not take the time to do that, and so what they read on the can doesn't really apply most of the time. Even though the can might tell you that it will shoot 10, 15, or 20 feet, usually you can count on about half of that. Pepper spray can also be relatively difficult to access and when someone sees you reaching into your pocket to get that out, they know what you are doing. They are smart, and they are going to keep you from getting to it.

Or, if you do spray someone and if they have been sprayed by pepper spray before, they can be conditioned to that by a large extent. They can still be getting their hands on you, hurting you, stabbing you, etc. plus they are also cross contaminating you with that pepper spray. The distance is so short with

Brett Lechtenberg

most of that stuff that it does not work out well in your favor.

So what you want to think about using is something like hornet spray, which has a really long reach. You want to be as far away from a hornet's nest as possible when you spray it, so this needs to shoot 20 or 25 feet. This spray will usually match what is on the can.

I recommend getting some of this spray from the hardware store, get a couple different brands and actually spray it and check it out. How far does it shoot? Find that brand. Bang, you have got a winner. What I like about hornet spray too is that it can sit on the back of a sink, could be in a bathroom, etc. and it blends into the background just like a screw driver does so people do not think much of it.

Now, say a home invader is walking / forcing you through your kitchen on the way to the basement – it is easy to grab that can of hornet spray and turn around to spray somebody from a distance and keep moving. There are a lot of advantages to that.

From a medical perspective, pepper spray eventually subsides and it can be washed off with soap and water to diminish the effects dramatically very quickly. Hornet spray, or something along that line, is poison. It is going to interfere with somebody's nervous system, their respiratory system, etc. and they have got to go to the doctor. They have a serious problem. They are not just going to wash that off.

Protecting Your Castle

Door Stops

Little rubber door stops that can be purchased for about 89 cents are awesome. Think about walking into a room, or into somebody's house, at some point in your past and the welcome mat got jammed up under the door, or the bath mat under the door in the bathroom and how hard that door was to open. Door stops are not going to keep someone out forever, but it is something they have to deal with that buys you time and creates an advantage for you. Rubber door stops also blend into the background, and they are really inexpensive.

Other Barriers and Weapons of Opportunity

There are several other things around your house that could be used as barriers or weapons of opportunity. For example, a heavy ashtray or a big heavy metal flashlight are things that you could hit somebody with. Stands for coats and hats could be pulled down behind you or in front of you as you are moving through a room. Moving your furniture, a high chair, anything like that could be used as a barrier to slow people down and buy you time. Those are really important.

Guns

I want to talk about guns for just a moment. I know we live in a gun culture. I do not have any problem with guns, I own some myself, and I am a believer. But I am only a believer if you are highly trained to use it. 80% of all first shots miss, and if you are in a home defense situation you do not have the luxury of missing. That bullet could easily travel through a wall and hit one of your family members when you do not hit your target.

If you have not trained in a tactical situation and trained under stress with these types of specific situations, then you probably do not want to be reaching for a gun in a high-stress home invasion situation. If you have trained for that, then that might be something that you could do.

If you are going to get a gun for home safety, then you need to think carefully about the caliber and other things that you get to make sure that you are making a good choice for your home and your family. I discuss all those things in my home invasion program and I am not going to go into all of that here.

Make good choices. Think about it in advance. Stage your home with weapons of opportunity.

Other Home Safety Tips

Here are a few bonus tips for extra security that are very rarely thought of around the house.

Something everybody that has outside window wells and a nearby hose spicket should be thinking about is sometimes people do random acts of violence and random acts of vandalism for no reason. There are people who will come by, grab your hose, stick it down in your window well, and turn on the water.

Protecting Your Castle

Your neighbors will not be any the wiser. If you were on vacation for a week and you come back – you have got a heck of a mess on your hands, probably a big insurance claim, and potentially tens of thousands of dollars of damage. This is an easy thing to prevent. So protect yourself, keep your family safe.

Next time you build your fence, repair your fence or replace it, think about how you can hide the handles and the hinges from the outside world so somebody looking at your house does not see a door in your fence as an easy access point. Your other neighbors are not going to think about these things and it will make their houses look easier for a burglar to approach and break into. It is okay to show the posts, but hide the hinges and handles so nobody can actually see the entry point.

Here is something else to think about as you go on vacation. A lot of times when people go on vacation they have someone pick up the mail. That is fabulous; continue to do that. However, they forget to think about things like pizza fliers and door hangers that can accumulate on their doors. Make sure you have got a neighbor or a trusted friend stopping by and picking those things up off your doors so it doesn't look like they are accumulating and showing that you are not home.

Another good idea is that if you have keys that have a key fob or some kind of panic button on them, instead of keeping them all in one central key location consider keeping at least one set in your bedroom near your bed. So if you think that someone has broken into your home or you have got some kind of problem, you can always push the panic button on your keys and set off the car alarm to help draw attention to yourself and maybe get some help a little quicker.

Because identity theft is on the rise and is one of the biggest crimes in the United States, it is a good idea to change out your mailbox and get one that has a lock and a key. That way when the mail comes, the mailman puts it into the mailbox and it drops down into an enclosed area where you have to have a key to get it out. Somebody walking by or driving by cannot open the door, grab your mail, have part of your identity, and then head on down the road. This will be a lot safer for you, your identity, and anybody else in your house.

Here is an important safety tip for anyone that has a home security system on where to place the keypad. Unfortunately, most people don't think about this very clearly. They will put their keypad as close to the main entrance as possible so they can turn it on and off very quickly and conveniently as they come in and out of their home. Unfortunately, many times that means it is visible from the outside through a nearby window or a window in a door, etc. for the average everyday burglar. A burglar will look at these things to see if your security system is active. If it is active, then they tend to move on. If it is not active, then that gives them the idea that they could potentially break into your home a lot easier. So

Protecting Your Castle

think about that when you place your keypad. Many times it can simply be placed around the corner from where it is and that makes your possessions and your family a whole lot safer.

SUMMARY

Concentric Rings of Protection

1. Fence line or outer yard

2. Main doors and windows

3. Seldom used rooms

4. Sleeping quarters

5. Safe rooms

Staging

- Weapons of opportunity

- Know your layout

- Do not panic

Protecting Your Castle

CHAPTER 5

What if You Find Someone in Your Home?

I cover anti home invasion tactics and home invaders extensively in my anti home invasion program, but it definitely warrants some discussion here.

Remember as discussed in the introduction, home invaders are a totally different type of human being. They are true predators. They do not value your life or your family's lives in any way. They simply look at you as something that is in their way or something that they can have their way with in order to satisfy any kind of deranged pleasures or fantasies that they may have.

Hostage Rooms

Let's talk about rooms in your home where a home invader may take you or your family to keep you from being seen or heard by the outside world. Things like an unfinished basement are areas that home invaders will use. Wherever there might be something that they could latch you to, to keep you there, so you can't escape is what they will look for. Things in finished basements where pipes may be coming in or out of the house, a toilet or heavy fixtures of some type are going to be things that they are going to be able to bind you to potentially to keep you from escaping.

Now they may not need any of those things. Home invaders generally come very well prepared with zip ties and rope and that kind of thing because they plan on being able to get you and put you somewhere and be able to secure you even if there is not something to keep you there like exposed beams, etc. So, I want you to think about the places in your home that are as far away from your street or neighbors as possible, that sound would have a hard time getting out of and nobody could really see into. Those are the places that where a home invader might stick you or your family.

Someone Is In Your Home

What do you do if you think someone has broken into your home? Maybe you heard a noise off in the distance, on the other side of the house, etc. and you think someone might be in your house. There are a couple of things you have to understand. You have to understand the concepts of what we are talking about and then the plan of how to deal with it.

So first, I will discuss the concepts. You need to

Protecting Your Castle

understand what is called the "Personal Safety Formula." The Personal Safety Formula is $E^3 + D$. The first E is Evaluate. The second E is Evade. The third E is Escape. And the D is Defend yourself if you must. It is important to understand it in that order because that will keep you the safest.

The second concept is "the situation dictates the response." There is no one correct answer for any given situation, however there will be certain kinds of specific responses you can use when given a particular *type* of situation. Most people have the huge misconception that there is some master technique or a master key principle that works for everything, and there is not. There are several keys to a lock that will open it, and some are better than others depending on the situation.

So think about it this way. Imagine a scenario where you are lying in bed. At the other end of the house you hear a noise. Maybe a window breaks or a door closes or opens that is not supposed to, etc. You have to immediately put the principles of $E^3 + D$, the Personal Safety Formula, into effect.

First, Evaluate what is going on. Where is it coming from? How are you going to get your family so you can Evade the problem? What are you going to do – are you going to Escape or are you going to bring yourself to a Safe Room? Are you going to have Defend yourself? You have to evaluate these things quickly. If you have been taking this training and you have been thinking about these things in advance,

43

this, although it will be scary, will not be that difficult. You will have a plan of action.

After you have exercised that $E^3 + D$ mentality, there are four master keys of what you need to put into play.

First, call 9-1-1 immediately. Make sure you get help on the way as quickly as possible. In another section I told you about criminal timelines and it is important that you make theirs as short as possible. So get the police or help on the way.

Part two is getting your family together. If that means going to your children's room and bringing them back to your room, or whatever your safe room is, great. If it means moving you and your spouse to your children's room and making that your safe room, that's fine. But you have to have thought about that in advance.

Now, depending on the situation you may want to go to a safe room that you have created in your house or you may want to get out. I cannot give you the answer based on everyone's home, where their kids are, where they sleep and how the house is set up. That is something you have got to decide - go to a safe room or get out. Both of those equate to Escape, because if you have set up your safe room properly you have a means of locking and blocking the door, you have disconnected yourself from the rest of the house, you are going to be able to get

Protecting Your Castle

more help by getting on the phone and you have essentially escaped the negative situation (at least as much you can) while being in your house.

Finally, "put the problem in spotlight." Now if you are trapped in your house, you are in your safe room, then you want to turn all the lights, you want to yell out the windows to your neighbors, you want to do whatever you can to make sure that when the police arrive they know where you are. These are just a few ideas, I cannot tell you that those will be right for every situation.

On the flip side of that, you may have decided to escape. That may have been the best plan, that's great. Still do everything you can to put things in spotlight. While you are getting out of the house maybe turn on as many lights as you can on the way out. Draw as much attention to the problem as you can. Maybe you are going to end up in your car, shining your headlights on the house, honking your horn. Putting it in spotlight doesn't necessarily just mean light, but bringing attention to the problem.

Those are some simple concepts and plans that you need to have thought out given the situation, the layout of the house, where your family is, etc. I cannot give you all the answers but I can give you the concepts to help you put together for your situation. In the world of protection the old saying is, "situation dictates response."

Tips on Confrontation

In this section I will tell you about what to do if you confront someone in your home. Maybe you are on your way to the bathroom and, surprise, there is somebody there. Or you heard a noise, you go to see if everything is okay with your children, and you come across somebody. You never know, right? I cover this extensively in my anti home invasion program, but I wanted to take at least a few minutes here and give you some common sense simple training that anyone can use.

The first thing you have to do is you have to give yourself permission to *do whatever it takes to keep yourself safe and your family safe*. Most people never do the mental shift of giving themselves that permission and they can end up in a moral dilemma wondering, "Should I hurt somebody? Should I not?" A lot of times people think, "Well, I'll just do it. I can do whatever it takes in the moment," but that is not reality. You are only going to be able to do in the moment what you have trained yourself to do and what you have given yourself permission to do. There is a huge misconception about what people can do in the moment and what they can't do in the moment without proper training – so give yourself permission.

Second, have a victor's mentality not a victim's mentality. This is really, really important. If you believe that you are going to be victorious, that you are going to be able to do what it takes and you are not going to back down when you are pushed into

Protecting Your Castle

that corner and have no choice, then many times will is way more important than skill. That has been proved time and time again throughout history.

Third, remember your advantages. If you come across someone in your house and they did not know that you were there, then you have the instant advantage of surprise which is a huge advantage. More battles and conflicts have been won because of the virtue of surprise than probably any other tactical advantage. That's awesome. You have that too.

In the event that you don't have surprise, remember that you have some other major advantages. You know the layout of your house. You know how to get to places and where things are that other people will not know. You know how to move around safely through your house. You know where barriers are – things like chairs, doors, children's toys, the couch, etc. that can be used as a barrier to slow someone down or potentially stop someone. You know where your weapons are in your house if you have some (gun, knife, bat, etc.). If you don't have any traditional weapons, then you know where all your improvised weapons are such as steak knives, heavy objects, etc. If you staged some weapons of opportunity you will know where those things are.

Finally, maximize funnels, or what is sometimes called a "Fatal Funnel." A hallway, a stairway, a doorway, and a window are all examples of a classic fatal funnel. They can cause things to go from a disadvantage to an advantage. If you had 2 or 3 people in your house and they were trying to get after you, then one of the things you could do is lead them into a hallway or through a doorway, where really only one person can go through at a time. This will help you reduce the threat of 2 or 3 down to 1. If you take advantage of that properly, you gain the advantage.

These are some examples of fatal funnels.

These are just some of the things that you can do if you confront someone in your house.

Protecting Your Castle

SUMMARY

What if Someone is in Your Home?

- Call 911 first

- Drive attackers into fatal funnels

- Stage potential hostage rooms

Brett Lechtenberg

CHAPTER 6

Wrapping it All Up

Thank you for reading and taking your family safety so seriously. Make your home safety a part of your everyday thoughts and habits.

Although encountering a true home invader is very rare, it is very possible that your home could be the target of a burglary or a robbery.

Even though you will never be able to totally stop someone from targeting your home or your family for a crime, there are definitely things that every person can do to help drastically minimize the chances of being the target of a criminal.

The key is to take your own and your family's safety and security seriously.

Most people just assume that everything will be fine and that nothing bad is going to happen to them. The reality is that burglary, just like identity theft, is very likely at some point in your life.

Do not live in denial. Face difficult situations head on.

If you want more information you can always go to my website www.brettlechtenberg.com and go to the Family Safety Resource Center where you will find over 60 free articles on family safety and security.

Protecting Your Castle

APPENDIX

In this appendix you will find the Quick Start Guide. Over the next several pages, I will give you examples of items mentioned in the program and even the links to purchase them online if that is easier for your household.

Brett Lechtenberg

Protecting Your Castle

A Common Sense Guide to Home Safety and Defense

Quick Start Guide

Thank you for purchasing *Protecting Your Castle.*

I know you will find this program loaded with helpful and simple tips. You will be able to implement the vast majority of this information almost immediately upon completing the program.

How to Maximize this Program and Get the Most out of it in the Shortest Amount of Time

- Watch the program in its entirety with the members of your household.

- I recommend anyone who is approximately 12 years old or older should watch this and understand the concepts in the program. In my opinion, there is nothing scary in this material.

- Purchase or organize the items I will talk about in the program, practice with them and place them in their appropriate places around your home.

This program goes over simple facts, tips and techniques on improving your home and personal safety. It is my opinion that the earlier people learn this type of material, the more it will become part of their way of life, thereby leading to a safer, less fearful and more empowered life.

Over the next several pages, I give you examples of items mentioned in the program and even the links to purchase them online if that is easier for your household.

Enjoy the program and be safe.

Items for Staging and Personal Protection

Disclaimer:

If you use anything as a weapon, then you could potentially open yourself up to a lawsuit. Also, in the heat of the moment someone could actually take a weapon from you and use it on you, turning a bad situation into a huge nightmare.

It is up to you as an individual to decide if you can actually use a weapon for your own or your family's protection and if you are actually willing to practice with this weapon in advance of it actually needing to be used.

*If you do not believe that you could use a weapon in a personal or family protection situation, and / or you are not willing to actually practice with the weapon, then by all means do **not** invest in these* *items. Also **never** believe that you will be able to do something in the heat of the moment that you have not trained for.*

There is an old saying that unfortunately people believe because they never heard the entire saying or chose not to believe the whole saying.

The saying is "People will rise to their level of expectation."

Unfortunately, many people have chosen to believe this when they have not heard or understood the entire saying.

*The actual saying is "People rise to their level of expectation, but **fall** to their level of training."*

If you do not train, you will most likely fall. The worst possible time to find this out for yourself is during a life and death situation for yourself or a family member.

Don't be so ignorant that you do not train.

Also I am in no way endorsing any specific product and I cannot tell you that they will not fail. You use any or all items, tactics, and techniques at your own risk.

Do not be a one trick pony. Be able to use anything as a weapon of opportunity should the need arise.

Protecting Your Castle

In the event that you cannot get to a hardware store for some reason, here are the links from Amazon for some suggested items. If you cannot get to a hardware store in the next couple of days then you might as well have them shipped to you. These items are not listed in any particular order.

Hornet Spray for Staging and Protection

This is my preference but if you like pepper spray or bear spray by all means go for it.

Your mileage may vary, but I have actually sprayed these to test their reach. I do **not** validate what is in the can.

Hot Shot Wasp and Hornet Killer

http://www.amazon.com/Hornet-Spray-Killer-Hot%20Shot/dp/B002Y6D2D0/ref=sr_1_2?ie=UTF8&qid=1439046755&sr=8-2&keywords=hornet+spray

Black Flag Wasp, Hornet and Yellow Jacket Killer

http://www.amazon.com/Black-Flag-Hornet-Aerosol-14-Ounce/dp/B00AA8WSSY/ref=sr_1_3ie=UTF8&qid=1439046755&sr=8-3&keywords=hornet+spray

Raid Wasp and Hornet

http://www.amazon.com/Raid-Hornet-14-Ounce-Discontinued-Manufacturer/dp/B006K3PZ3S/ref=sr_1_5?ie=UTF8&qid=1439046755&sr=8-5&keywords=hornet+spray

Protecting Your Castle

Screw Drivers for Staging and Protection

I have generally found that buying a 2 pack is usually the best way to go because you get a good value and screw drivers that are of adequate length. I would avoid the mega packs unless you need screw drivers of varying lengths for home projects.

Stanley 2 Pack

This pack works out to be $1 per screw driver. They have usable handles and are a good length. I personally have no brand preference.

http://www.amazon.com/Stanley-60-020-2-Piece-Standard-Screwdriver/dp/B000FK6ZN6/ref=sr_1_6?ie=UTF8&qid=1439048120&sr=8-6&keywords=screw+drivers

Door Stops for Staging and Protection:

I recommend keeping a door stop by every main door entrance to your home and in each bedroom and your safe room. Make sure to practice positioning the stop with your foot under the door as you close it so you know you can do these motions under stress.

Rubber Door Stops

http://www.amazon.com/SuperiorMaker-Surfaces-Including-Carpet-Securely/dp/B00MC4NL7G/ref=sr_1_1?ie=UTF8&qid=1439048938&sr=8-1&keywords=Rubber+door+stops

Protecting Your Castle

Door Stop with a Buzzer

These are more expensive but they do make noise that could scare someone away.

http://www.amazon.com/Skque-Wireless-Security-Warning-Stopper/dp/B00974VLMQ/ref=sr_1_1?ie=UTF8&qid=1439049079&sr=8-1&keywords=door+stops+with+buzzer

Medical Kit for Your Safe Room

Although a big trauma bag may be a bit of overkill, I always believe that it is better to have medical supplies and not need them than to need them and not have them. If one person in your house gets hurt one time, then you will never be sorry you had plenty of medical supplies.

http://www.amazon.com/First-Emergency-Response-Trauma-Complete/dp/B00HQ15XKI/ref=sr_1_5?ie=UTF8&qid=1439049923&sr=8-5&keywords=medical+kit

Protecting Your Castle

REFERENCES

1. www.BrettLechtenberg.com

2. www.Amazon.com

Brett Lechtenberg

Protecting Your Castle

ABOUT THE AUTHOR

Brett Lechtenberg is a nationally recognized expert on personal and family safety. Brett has dedicated his life to educating people about family safety and personal security. Through his comprehensive lecture series, motivational speaking and personal protection courses, Brett has trained thousands of everyday people to empower themselves both physically and mentally as they learn the safest and most effective methods of personal and family protection.

Brett's first book "The Anti-Bully Program" became an Amazon #1 Best Seller in three categories, has appeared on Good Day Utah, Channel 13 news broadcasts and Channel 4 news broadcasts. Brett has been a speaker and presenter for local chambers of commerce groups, local schools, local Exchange Clubs, Draper and Salt Lake City code enforcement officers, the Arizona State Process Servers Association, Delta Airlines, American Express, Citigroup and many more local community groups and churches.

You can contact Brett directly through his website at:
http:// brettlechtenberg.com/

Follow Brett on Facebook: https://www.facebook.com/pages/
BrettLechtenberg/1413417442206269

Follow Brett on youtube:
http://www.youtube.com/user/brettl99

Printed in Great Britain
by Amazon